A COMET CALLED

Ian Ridpath and Terence Mur

GW00360236

The right of the University of Cambridge to print and sell all manner of books was granted by Henry VIII in 1534. The University has printed and published continuously since 1584.

Cambridge University Press

CAMBRIDGE
LONDON   NEW YORK   NEW ROCHELLE
MELBOURNE   SYDNEY

Published by the Press Syndicate of the University of Cambridge
The Pitt Building, Trumpington Street, Cambridge CB2 1RP
32 East 57th Street, New York, NY 10022, USA
10 Stamford Road, Oakleigh, Melbourne 3166, Australia

© Cambridge University Press 1985

First published 1985

Typeset by Glyn Davies, Cambridge
Printed in Great Britain by Blantyre Printing

British Library cataloguing in publication data

Ridpath, Ian
   A comet called Halley.
   1. Halley's comet
   I. Title   II. Murtagh, Terence
   523.6′4   QB723.H2

Library of Congress cataloguing in publication data

Ridpath, Ian A Comet Called Halley
   I. Halley's Comet
   I. Murtagh, Terence
   II. Title QB723. H2R53 1986 523.6; 4 85-18981

ISBN 0-521-31282-5 (pbk)

CONTENTS

Comets can appear from any direction in space. Studies of their orbits suggest that they all come from a great swarm of cometary debris which may lie halfway to the nearest star.

Attracted by the immense gravity of the Sun, comets sweep through the Solar System in a variety of graceful elliptical orbits. Not all survive their visit to our star. Some plunge directly into it and are destroyed. Others, their paths disturbed by a close approach to a planet, have their orbits changed and are captured forever to orbit the Sun. Most, however, escape back into the depths of space from which they came.

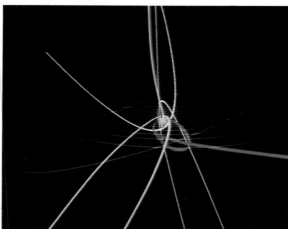

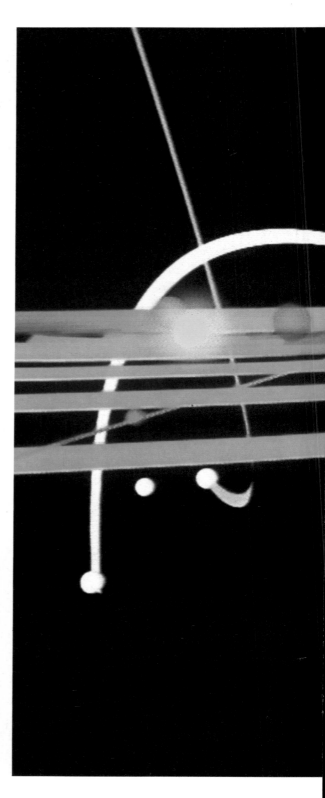

The Comet arrives

Far from the Sun, a comet resembles a snowball of ice and dust a few kilometres across.

About 170 000 years ago a comet approached the Sun from the darkness of space. As it crossed the orbit of Pluto, the outermost planet of our Solar System, the comet, still 5900 million kilometres distant, shone only by reflecting the feeble light of the Sun. Out there, the comet's temperature was below −150°C. At that stage, the comet resembled a dirty iceberg of dust and rock, cemented together by frozen gas. It was about 20 kilometres across – on the large side as comets go, but not exceptional.

Approaching closer to the Sun's fire, the comet began to warm up. Somewhere between the orbits of Jupiter and Mars, about 500 million kilometres from the Sun, the ice of the comet began to evaporate, enveloping the central iceberg in a fuzzy halo that fluoresced like a neon tube. Gas and dust streamed away from the comet's head to produce a tail. Never before had the comet experienced warmth or light, and in these congenial conditions it blossomed.

Continuing its quickening plunge towards the Sun, the comet made its first ghostly appearance in the skies of Earth to prehistoric man, early forms of *Homo sapiens*. When darkness fell, who among them was the first to see the ethereal glow of the unexpected visitor? Silently gliding in front of the stars from night to night, the celestial spectre filled its primitive onlookers with dread, as it would future generations in its appearances down the ages.

After passing closest to the Sun, the comet shone more gloriously than ever. Its head had swelled to 100 000 kilometres across, and its tail issued like a cosmic ectoplasm for 100 million kilometres and beyond.

But its glory was not to last. Receding from the Sun again, the comet wilted. Its tail shortened and vanished, and its head shrank back into its icy nucleus. Only a few months after it first flowered, the comet was heading back into the deep-freeze of space whence it had come, to the invisible cloud of comets ringing the Solar System.

It might never have been seen again but for a chance encounter with the giant planet Jupiter, whose gravitational embrace swung the comet onto a new and shorter path that has brought it back to the Sun regularly ever since. We now call it Halley's Comet in honour of Edmond Halley, the English astronomer who calculated its orbit in 1705 and who thereby established that comets are members of the Solar System, subject to the same laws of gravity as govern the motions of the planets.

Counting comets

While Halley's Comet is the best-known comet, and has been seen more often in history than any other, there are plenty more of these celestial wanderers to entertain astronomers. As of May 1985, a total of 735 comets had been followed carefully enough for their orbits to be calculated, according to the chief accounting office of comets, the Harvard–Smithsonian Center for Astrophysics in Cambridge, Massachusetts. No one can tell how many more have been sighted briefly and subsequently lost. Each year the number of known comets grows. In 1983, the record year to date, a total of 22 comets was seen. Of these, 12 were new discoveries, while the remainder were known comets returning.

Even outside record years, the numbers seen are not much less – for example, 21 in 1984. Most of these are faint, though, requiring large telescopes to view them. Only occasionally does a comet become bright enough to be easily visible to the naked eye, such as Comet Ikeya–Seki in 1965, Comet Bennett in 1970, Comet West in 1976 – and of course Comet Halley in 1986.

Comets fall in towards the Sun from a great cloud, way beyond the visible planets, known as the Oort Cloud after the Dutch astronomer Jan Oort who proposed its existence in 1950. Left to its own devices, a comet would swing around the Sun and head back out again into the Oort Cloud. But in practice, a comet's path is usually tweaked by the gravitational pull of the planets, notably Jupiter. Sometimes the tweak can throw it out of the Solar System altogether, but more often it is reined into a tighter loop.

Astronomers distinguish two main cometary groups: the long-period comets, which take many thousands or millions of years to orbit the Sun; and the short-period comets, sometimes known simply as periodic comets, whose orbits have been tweaked so often by close passages to Jupiter that they now lap the Sun in less than 200 years. (The dividing line is somewhat arbitrary, chosen because there are virtually no comets with periods of around 200 years.) Halley's Comet is the only really bright example of the periodic class. In the Harvard–Smithsonian accounts, 603 long-period comets are listed, fully 82 per cent of the total; the remaining 132 (18 per cent) are short-period comets.

One particularly daredevil family of comets is the Sungrazers, so named because they virtually scrape the Sun at their closest approach. One such was Comet Ikeya–Seki, which in 1965 passed a mere 500 000 kilometres from the Sun's incandescent surface. This distance, only slightly more than the separation between the Earth and the Moon, is a hair's breadth in astronomical terms. Sungrazing comets have orbits that bear a distinct family resemblance to each other, and one theory is that they are the fragments of a much larger comet that split after a succession of close approaches to the Sun. Some Sungrazers fail to emerge from their run of the gauntlet; they crash into the Sun.

Comet Burnham, as photographed on April 27 1960 by Hendrie and Ridley.

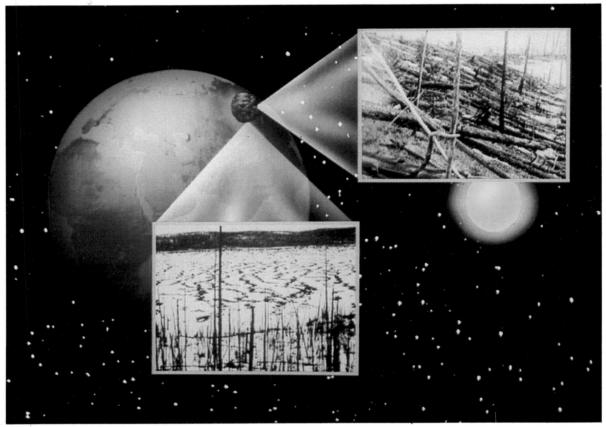

In 1908 a fragment of a comet slammed into Siberia, flattening trees like matchsticks for up to 30 kilometres around.

Collisions with comets

With so many comets jaywalking across the paths of the planets, it is not surprising that traffic accidents occur. Earlier this century a Siberian farmer, S.B. Semenov, had a closer view of one comet than he would have wished.

While sitting on his porch on the morning of June 30, 1908, at the remote trading station of Vanovara, Semenov suddenly saw a huge fireball like a second Sun blazing earthwards about 60 kilometres away. Semenov felt its intense heat scorching his skin. A few seconds later the blinding fireball exploded above the taiga with a force that knocked Semenov off his porch. Deafening thunderclaps rent the air, and the ground shook.

Nearer the centre of the blast, in the valley of the Stony Tunguska River, trees were felled like matchsticks for up to 30 kilometres around. Heat from the explosion melted metal objects, burned reindeer to death, and set the forest ablaze. Miraculously, no people were killed. Six hundred kilometres to the south, near Kansk, the driver of the Trans-Siberian Express brought his train to a screeching halt for fear of a derailment as the ground shuddered. Shock waves like those from an earthquake were recorded throughout Europe. In England on the night of July 1 the sky glowed mysteriously after midnight, a result of dust thrown into the upper atmosphere. It was a natural disaster rivalled only by the eruption of Krakatoa.

The Tunguska blast was the rare encounter of the Earth with a comet – fortunately a small one, perhaps 100 metres across and weighing a million tons. Ripped apart by the forces of deceleration at a height of 8 kilometres above the Earth, the comet's crumbly head shattered explosively with the force of a 12.5-megaton bomb, laying waste the taiga below with heat and shock waves. Dust from the disintegrated head of the comet has been found in the soil of the area; its composition matches that of cosmic dust collected from the upper atmosphere by rockets.

When it struck Siberia, the comet was heading away from the Sun, which allowed it to sneak up on us unnoticed. But by backtracking its direction of approach, astronomers have found that it was on an orbit similar to that of Comet Encke, the comet with the shortest known period, 3.3 years. According to Czech astronomer Lubor Kresak, the Tunguska object was probably a fragment that broke away from Comet Encke several thousand years ago.

Beware, say the astronomers. Objects the size of the Tunguska comet are due to hit the Earth every 2000 years or so, on average. Next time, more than a few Siberian reindeer could be fried. To protect the Earth, we may need to set up a space-based battery of nuclear missiles, turning the Solar System into a cosmic shooting gallery against real-life space invaders.

The death of the dinosaurs 65 million years ago may have been caused by the impact of a comet with the Earth. Dust from the impact, thrown high into the atmosphere, would have blocked out light from the Sun for months or years. Plants and animals would have perished in the gloom.

Did a comet kill the dinosaurs?

Larger objects than the Tunguska comet hit the Earth less frequently, because there are fewer of them, but given time, they too, must arrive. Sixty-five million years ago, such an event may have happened. Then, an environmental disaster wiped out half the families of living things on Earth. Notable among the casualties were the dinosaurs, but other land and sea creatures were extinguished with them.

Marking this event in the geological record is a layer of clay, a centimetre thick, whose composition is not of this Earth. Specifically, it contains too much of the rare metal iridium, plus other metals such as platinum and gold, to be terrestrial, but this composition closely matches that of meteors and meteorites. Below the thin layer of clay lie the rocks of the Cretaceous period, including dinosaur bones. Above the clay lie the rocks of the Tertiary period, with no dinosaurs.

If the theories are right, this clay contains the ashes of an object that put an end to the dinosaurs when it slammed into the Earth 65 million years ago. Estimates put the size of this object at about 10 kilometres across, similar to the nucleus of a large comet such as Halley's. Everything within a few hundred kilometres of its impact point was pulverized in the blast, equivalent to the explosion of a nuclear bomb of millions of megatons.

Any dinosaurs unfortunate enough to be standing 2000 kilometres or so from the impact had several minutes to ponder the remarkable mushroom cloud of dust rising on the horizon before its effects reached them. They were incinerated by a blast of air at 500°C, and winds of 2400 kilometres per hour scattered their charred remains far and wide. Where they had stood, the ground heaved with movements as high as a house. Overhead, a stream of hot ash fell as fiery rain. Even 10 000 kilometres away, hurricane-force winds blew devastatingly for over 12 hours.

In addition to those immediate effects, the extra-terrestrial impact left a legacy like a nuclear winter. Dust pumped high into the atmosphere shrouded the entire Earth from sunlight for several years. In the darkness, plants rapidly wilted and died, starving the remaining dinosaurs and other animals which relied on them for food. By the time the dust began to settle and the Sun dimly reappeared, it was too late. The last of the dinosaurs was dead. Over the bones settled the dust of the object that had killed them; this dust subsequently formed the layer of clay.

Geologists have found signs of other mass extinctions in the fossil record. According to the catastrophists, waves of comets may sweep through the Solar System at intervals. Their impacts may have fundamentally affected the geological and biological history of our planet. Maybe the ancients were right to fear comets after all.

Comet lore

A bright comet is an awesome sight. The best of them resemble glowing tufts of cotton wool, unfurling their gossamer tails so far across the sky that your outstretched hand could not cover them. Unlike shooting stars, which suddenly flash out in the fashion of a sky rocket, comets hover serenely against the background of stars, like a finger admonishing the inhabitants of Earth below. A comet's flowing tail gives an impression of a headlong dash that is entirely illusory. In reality, its movement against the stars is noticeable to the naked eye only over a period of hours, or from night to night.

A comet *looks* like a portent, and it is not surprising that people have always regarded them as such. Writing 2000 years ago, the Roman astrologer Manilius summed up the prevailing opinion: 'Heaven in pity is sending upon Earth tokens of impending doom'. Included in his list of cometary ills were blighted crops, plague, wars, insurrection, and even family feuds. In short, anything could be blamed on comets, and usually was.

Similar beliefs prospered in the Far East, where comets were called 'broom stars'. From a Chinese tomb of 168 BC, archaeologists have uncovered a set of paintings on silk which amount to an identification guide to various forms of comet tails and the events they were said to foreshadow, including wars, famine and death.

Not everyone took an unremittingly gloomy view. In AD 77, the Roman writer Pliny, in his book *Natural History*, gave his version of how to interpret a comet: 'If it resembles a pair of flutes, it is a portent for the art of music; in the private parts of the constellations it portends immorality; in relation to certain fixed stars it portends men of genius and a revival of learning; in the head of the northern or the southern serpent (i.e. Draco and Serpens) it brings poisonings.'

In the West, beliefs about comets were influenced for more than 2000 years by the Greek philosopher Aristotle, who declared in the fourth century BC that comets were strictly atmospheric phenomena. In Aristotle's cosmology, the Earth was stationary at the centre of the Universe, and all celestial bodies – the Sun, Moon, planets and stars – revolved around the Earth on spheres of pure crystal. Nothing could be allowed to violate the perfection of the heavens, so that any temporary blemish such as a comet had to be assigned to the atmosphere. According to Aristotle, comets were produced by gases that rose into the upper atmosphere where they caught fire, apparently being ignited by sparks generated by the motion of the heavens around the Earth. If the gases burned quickly, they produced the sudden flash of a shooting star. If they burned slowly, a comet was the result.

In the second century AD, the Greek astronomer Ptolemy reported in his *Tetrabiblos* that comets contained everything you needed to make a detailed prognostication, provided you knew how to read the signs aright: 'They show, through the parts of the Zodiac in which their heads appear and through the directions in which their tails point, the regions upon which the misfortunes impend. Through the formations of their heads they indicate the kind of the event and the class upon which the misfortune will take effect; through the time which they last, the duration of the events; and through their position relative to the Sun likewise their beginning.' Similar claims were still being made in the Middle Ages, and such beliefs have not entirely died out even today.

Of all the ancient writers on comets, the one to emerge with most credit is Lucius Seneca, a Roman of the first century AD. Seneca contested Aristotle's view that comets were sudden fires, arguing instead that they were celestial bodies moving on orbits like planets and that they might reappear, given time. Prophetically he wrote: 'Men will some day be able to demonstrate in what regions comets have their paths, why they move so far from the planets, what is their size and constitution'.

But in his day Seneca was ignored as a killjoy out to ruin a good business for the soothsayers. Seventeen centuries elapsed before the first part of his prediction came true (the second part, concerning the size and constitution of comets, is being fulfilled now). In the meantime, the art of cometary prophecy flourished, encouraged by the Church which was pleased to regard comets as signs from God. You could read into a comet whatever your imagination fancied. Comets were the UFOs of their day.

Comet West had a complex tail consisting of broad streamers of dust and a straighter, brighter tail of gas.

Aristotle, the ancient Greek scientist, thought that comets were luminous gases in the Earth's atmosphere. His views were accepted for 2000 years.

By comparing observations of the comet of 1577 made from two places at the same time, Tycho Brahe showed that the comet was not in the Earth's atmosphere but lay further away than the Moon.

Understanding comets

Comet fever reached new heights in the fifteenth and sixteenth centuries, when a total of 21 comets was seen. Ever alert, prophets of doom began to churn out lurid pamphlets predicting all manner of associated evils. They have been at it ever since.

But this was also the time when cometary science began to take tentative steps forward. One of the contributory factors was Halley's Comet, although no one knew it by that name at that time. A German astronomer, Peter Apian, observed Halley's Comet in 1531 and reported that its tail always pointed away from the Sun. Apian's observations, printed in a book with beautiful hand-coloured drawings, made a great impression. Comet tails do indeed flee from the Sun, no matter in which direction the comet is travelling, but the full explanation had to await twentieth-century physics.

In 1577 Tycho Brahe, the greatest observer of the pretelescopic era, made a breakthrough that was literally shattering. From his observatory on the Danish island of Ven, Tycho demonstrated that the bright comet of 1577 lay far beyond the Moon and in the realm of the planets, in contradiction to the teachings of Aristotle.

Tycho was not a man to worry about stepping on the toes of authority. As he watched the Comet take its course through the heavens from night to night, somewhere in his imagination he must have heard the sound of breaking glass – the crystal spheres of Aristotle that the comet had shattered. It was an impressive example of observation overthrowing a theory.

But what paths did comets follow through space? It would have helped if astronomers had known how the planets moved. At that time there was still confusion over whether the Sun and planets went around the Earth, as in the old system of Aristotle, or the Earth and planets around the Sun, the new theory due to Copernicus. In 1609 the German mathematician Johannes Kepler settled the matter by calculating, from the observations of Tycho Brahe, that the planets orbited the Sun along elliptical paths.

Kepler also turned his attention to comets following the appearance of a comet in 1607 (actually, Halley's Comet again). Curiously, Kepler considered that comets moved through the Solar System in straight lines, though in fairness the observations available to him were insufficient to compute an accurate orbit. Despite this lapse, he had some astute ideas about the formation of comet tails: 'The direct rays of the Sun strike upon it [the comet], penetrate its substance, draw away with them a portion of this

Tycho Brahe's discovery that comets moved among the plants shattered the theory that planets revolved on crystal spheres.

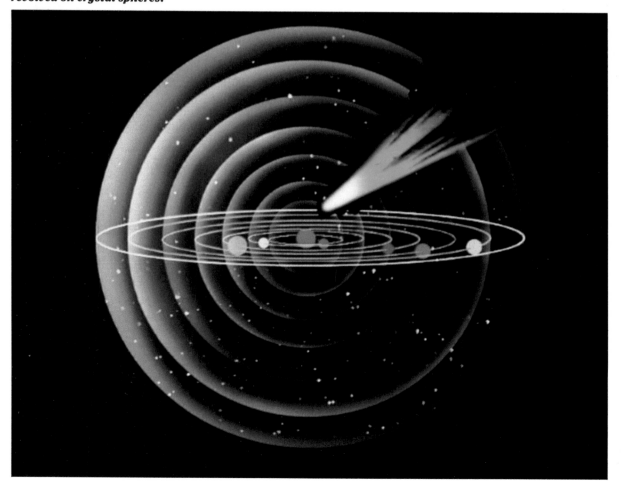

Johannes Kepler, the great German mathematician, thought that comets moved in straight lines. He was wrong.

matter, and issue thence to form the track of light we call the tail . . . In this manner the comet is consumed by breathing out its own tail.' Comets, he surmised correctly, are as numerous as fish in the sea, but we see only a selection of them.

In 1664 and 1665 two bright comets appeared, and between them an eclipse of the Moon was seen. Such a triple omen was unique. One can almost hear the intake of breath in anticipation of the unparalleled disasters that surely must follow. Lest anyone be uncertain about the meaning of these omens, John Gadbury, an English astrologer, thoughtfully interpreted them in his book of 1665, *De Cometis*. 'These Blazeing Starrs! Threaten the World with Famine, Plague, & Warrs', he trumpeted. 'To Princes, Death: to Kingdoms, many Crises: to all Estates, inevitable Losses!' He can hardly have believed his luck when London was hit by the Black Death in 1665 followed by the Great Fire the year after. Unwittingly, he had demonstrated a fact that modern-day astrologers know well: the laws of chance ensure that you can't be wrong *all* the time.

While London suffered, in Danzig one of the greatest astronomers of the day, Johannes Hevelius, was watching the comets with scientific detachment. He published his observations in 1668 in a volume entitled *Cometographia*, in which he theorized that comets are thrown out by the planets, notably Jupiter and Saturn, and move past the Sun on boomerang-shaped curves. Unlike boomerangs, though, they never came back. One attractive feature of Hevelius' book is a series of drawings of comets. To the untutored eye they may resemble the amputated tails of small furry mammals, but they were the most accurate renditions up to that time.

A number of astronomers began to suspect that comets orbited the Sun on paths like exaggerated forms of the planets' orbits, but no one could prove it. The time was ripe for the emergence of a man who, in the words of Seneca, could tell the paths of the comets. In fact, it took not one man but two: Isaac Newton and Edmond Halley.

Johannes Hevelius published a book in 1668 containing these drawings which show how a comet's tail points away from the Sun.

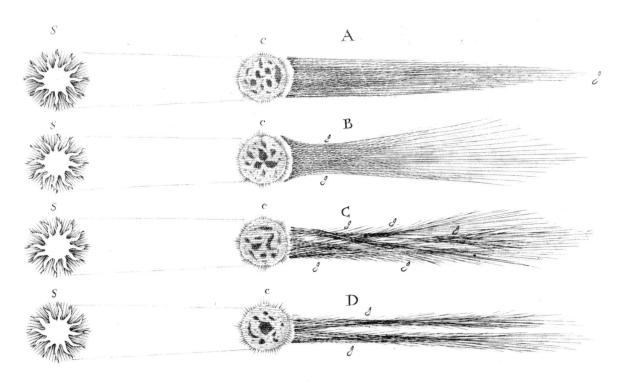

Sun

Edmond Halley

Isaac Newton and Edmond Halley studied the path of the comet of 1680. Newton published this diagram of the comet's orbit in his book 'Principia'.

Edmond Halley was perhaps the second-greatest genius in the history of British astronomy. By chance, he lived at the same time as the greatest genius of all, Isaac Newton. In one sense this was unfortunate, for Halley's reputation has been overshadowed by the stature of Newton. But in another sense it was a happy chance indeed, for had the two not been contemporaries, Halley's Comet would most likely be known by a different name.

Halley was born in 1656 in Haggerston, then a village north-east of London but now engulfed in the urban sprawl of Hackney. Halley's father manufactured soap. Why young Edmond turned to astronomy is uncertain, but the comets of 1664 and 1665 may well have played a part. Whatever the reason, his interest in astronomy was already well developed when he went to Oxford University in 1673, where he proved himself a brilliant student. He soon came to the attention of the newly appointed Astronomer Royal, John Flamsteed, with whom Halley observed at Greenwich as a vacation student.

Halley possessed self-confidence and ambition equal to his intellect. Without waiting to finish his studies at Oxford, he persuaded his father to support him on a two-year trip to the island of St Helena in the South Atlantic, from where he planned to catalogue the southern stars as Flamsteed was doing for those in the north. Coming from any other 19-year-old, it would have sounded embarrassingly presumptuous. Halley returned in 1678 with the job accomplished. He was rewarded with the accolades of the Royal Society and its patron, King Charles II, who ordered Oxford University to award Halley his master's degree.

16

Halley and his Comet

Edmond Halley's table of cometary orbits showed the similarity of the comets of 1531, 1607 and 1682.

Preserved in the archives of the Royal Greenwich Observatory is the very notebook in which Halley wrote the sightings of 'his' comet. It consists of 180 pages, 7¼ inches deep by 5¾ inches wide, now brown and stained by age, containing a jumble of calculations, geometrical figures and observations in Halley's often untidy handwriting. Originally it was a college exercise book which Halley continued to use for many years after he left University, gradually filling in all the blank spaces with notes in English and Latin (the scientific language of the day), sometimes upside down and even over the top of previous entries.

Evidently Halley did not yet have his telescopic sextant set up at Islington, for he made only rough naked-eye estimates of the Comet's position — witness this entry for August 31: 'The comet was on the straight line which passed through the leading shin of Boötes and the elbow of his left arm. Then clouds near the horizon obscured the comet.' When Halley came to calculate the orbit of the Comet 13 years later he never used his own observations but relied instead on the precise positions obtained by Flamsteed at Greenwich.

Still puzzled by the wayward motions of comets, Halley in 1684 travelled to Cambridge to consult the greatest mathematician of the age, Isaac Newton. There Halley found to his amazement that Newton, in scholarly isolation, had been at work for nearly 20 years on a theory of gravity that would explain the orbits of the planets and of comets. Halley urged that Newton should write it up for publication and he undertook that the Royal Society, of which he was now Clerk, would publish it.

Thus began the story of the *Principia*, one of the greatest scientific books ever written, which presented to the world Newton's theory of gravity and his laws of motion. As events turned out, the Royal Society pleaded penury and Halley, the only person who understood the full significance of the book, ended up paying for its publication himself. Its first edition came out in 1687. If Newton ever offered to help with the expenses, that fact is not recorded.

Comets feature prominently in the *Principia* as proof that the Sun's gravity controls the paths of everything within its ambit, not just the planets. Straight lines like those of Kepler were no longer conceivable: a curved orbit was required. Based on the motion of the

Significantly, a comet was in the morning sky at the end of 1680 when Halley left on a trip to Paris. By the time he arrived in Paris the comet had rounded the Sun and reappeared impressively in the evening sky. It was the talk of the city but it was not generally realized to be the same comet seen a few weeks earlier. Even the great Isaac Newton, watching from Cambridge, thought there had been two separate comets. Halley tried to calculate the comet's motion, taking Kepler's word that such a body should travel in a straight line. Not surprisingly, he failed. The matter rankled with him, for he was not a man to be easily beaten.

Back in London in 1682 Halley married and, as if by an act of Providence, another comet appeared. This was the comet that would one day bear his name, though Halley could never have guessed that fact as he scribbled down observations of it at his new home in Islington.

Two pages from Edmond Halley's notebook containing his observations, written in Latin, of the comet of 1682, which we now call Halley's Comet.

comet of 1680 Newton concluded that the right shape was a parabola, which is a curve like a long, thin ellipse but which never closes in on itself.

Gladly seizing the hint, Halley embarked on a long-term study of comets and their orbits. He came to realize that the orbits of most comets, perhaps of all, were not parabolic but were elliptical in shape, like exaggerated versions of the planets' orbits. Some years later, while checking some figures, he found that one of the positions Newton had used for the comet of 1680 was incorrect. Substituting the correct figure, Halley was able to calculate a better orbit. It was an ellipse after all.

Halley was laboriously building up a table of orbital statistics against which any 'new' comets could be compared to see if they had appeared before. Consequently he was particularly excited by the comet of 1682, the one he had witnessed from Islington. If his calculations were correct – and since he was using the excellent measurements made by Flamsteed at Greenwich there was no reason to doubt it – this comet was the same one that had been seen by Kepler in 1607 and by Apian in 1531, orbiting the Sun every 75 years or so. Slight differences in the interval between one appearance and the next could be accounted for by the gravitational nudges of the planets.

The fruits of his labours were published in 1705 in a paper entitled 'Synopsis of Cometary Astronomy'. More than 10 years of wearying research and computation were distilled into a table listing 24 comets and the vital statistics of their orbits. Halley declared his suspicion that the comets of 1531, 1607 and 1682 were the same, and ventured 'to advise posterity carefully to watch for its return about the expected year 1758'. As time went by, his confidence in his prediction grew. In a revised version of his paper, published posthumously, he went so far as to make a request of history: 'If according to what we have already said it should return again about the year 1758, candid posterity will not refuse to acknowledge that this was first discovered by an Englishman'.

Halley's remaining career was a glorious one. In 1720 he was appointed to the post of highest distinction in his profession, that of Astronomer Royal. Although he was then aged 63, his energy was undiminished. He embarked on an observing programme of the Moon's motions that lasted 22 years until his death in 1742 at the age of 85. It was then 16 years before his Comet was due to reappear.

The Comet returns

Halley was aware that his prediction of the Comet's return could be in error by many months, for he had only roughly estimated the effect of Jupiter on the Comet's path and had not taken the effect of Saturn into account at all. Two French astronomers, Alexis Clairaut and Joseph Lalande, spent several arduous months refining the prediction. They found that the actions of Jupiter and Saturn would delay the Comet's arrival at perihelion (closest to the Sun) until April 1759.

So great was the anticipation that scarcely a night can have passed in 1758 without some astronomer in Europe looking out for the Comet. As the months wore on with no sighting, doubts began to surface. At last, on Christmas night, 1758, a German amateur astronomer, Johann Palitzsch, observing with a home-made telescope, saw the smudgy image of the Comet. It reached perihelion on March 13, 1759, a month earlier than the revised date calculated by Clairaut and Lalande. (Had they known of the existence of the planets Uranus and Neptune, not yet discovered, their prediction would have been even more accurate.)

Now there could be no doubt. Comets *were* periodic and they *were* members of the Solar System, subject to the same laws that guided the planets in their paths. And Edmond Halley, guided by the hand of Isaac Newton, was the man who had proved it. Halley's Comet is his permanent memorial, looping around the Sun once in a human lifespan.

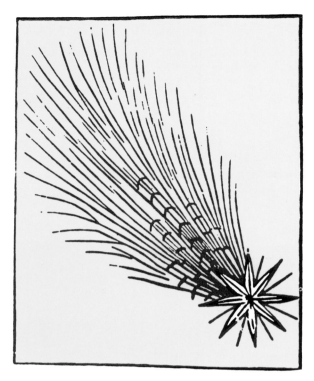

A woodcut from the 'Nuremberg Chronicles' showing Halley's Comet at its appearance in AD 684.

Saying hallo to Halley

Traditionally the name Halley has been pronounced to rhyme with valley, as in this Royal Astronomical Society drinking song from 1910:

Of all the comets in the sky,
There's none like Comet Halley.
We see it with the naked eye
And periodically.
The first to see it was not he,
But still we call it Halley.
The notion that it would return
Was his originally.

Most people nowadays say 'Hailey', a legacy of the rock'n'roll days of Bill Haley and the Comets. To add to the confusion, some historians have suggested that Halley might have pronounced his own name 'Hawley'. Unfortunately, there are no known Halleys descended from the great astronomer to help us.

We telephoned representatives of 36 Halley families and asked them how they pronounced their own name: two said it 'Hailey', and two others said 'Hawley'. The rest, 89 per cent of our sample, used the traditional pronunciation, which is also used by most astronomers engaged in comet study. Whichever way Halley himself pronounced his name, the traditional pronunciation clearly has the support of the vast majority of present-day Halleys.

The history of Comet Halley

Since its first predicted return in 1759, Halley's Comet has come back to us three times, in 1835, 1910 and now in 1985–86. Historical records, unknown to Edmond Halley, show that the Comet's previous appearances stretch far back into the past, long before comets were recognized to be periodic. Including its current visit, Halley's Comet has been sighted on 30 occasions, the earliest by the Chinese in 240 BC, and there may be still older records that remain undiscovered.

To chase the orbit of a comet back through time is not such an easy task as it might sound, for the gravitational effects of the planets are perpetually tugging on the comet and modifying its orbit. If the comet's orbit is slightly expanded by the pull of the planets, it will take longer to return; if the orbit shrinks, the comet will come back sooner. Powerful computers are needed to enumerate the changes, and the calculations must be checked regularly against actual observations of the comet.

Prior to its appearance in 1456, the best observations of Halley's Comet come from China. For over 2000 years, the Chinese government ran an astronomical bureau whose officials kept a scrupulous watch on the sky, noting anything unusual and interpreting its portents for the Emperors. Their records are an encyclopaedic source of information about ancient events in the sky.

Of the present-day astronomers who have tried backtracking Halley's Comet, the most successful have been Donald Yeomans, who uses NASA's computers at the Jet Propulsion Laboratory in California, and Tao Kiang, a Chinese astronomer based at Dunsink Observatory in Ireland. Yeomans and Kiang have calculated that the Comet made an exceptionally close passage to the Earth in AD 837, a mere 5 million kilometres away, and there are extensive descriptions of this spectacular 'broom star' from China: 'On the night of April 9 its length was more than 50 degrees. It branched into two tails . . . On the night of April 11 the length of the broom was 60 degrees. The tail was without branches and it pointed north. The Emperor summoned the Astronomer Royal and asked him the reason for these star changes.' One hopes that the Astronomer Royal had a good answer handy, for tradition has it that two Chinese astronomers who were caught napping by an unexpected eclipse were beheaded.

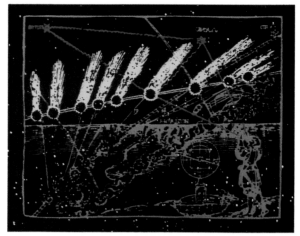

Peter Apian's observations of the comet of 1531 showed that its tail always pointed away from the Sun.

Yeomans and Kiang found that the earliest reliable record of Halley's Comet was in 240 BC, when a Chinese source briefly noted: 'During this year a broom star was seen at the north direction and then at the west direction'. Then comes the inevitable genuflection to superstition: 'During the summer the Empress Dowager died'.

Surprisingly, the Chinese did not record the Comet's subsequent appearance in 164 BC, and only vaguely mentioned it in 87 BC. This was vexing for Yeomans and Kiang, who needed to check their calculations for those years. But confirmation came from a different direction: the Babylonians, inhabitants of the Middle East, who assiduously compiled diaries of astronomical information. Their annals for 164 BC and 87 BC, recently decoded, contain unmistakable references to a comet that can only be Halley's, putting in its regular appearances according to the schedule calculated by Yeomans and Kiang. It is a remarkable fact that observations made with the naked eye 2000 years ago, and impressed into clay tablets in cuneiform script, are still of value in this day of high-speed computers and space probes. Perhaps even older records of the Comet, still unnoticed, lie etched on the Babylonian clay tablets stored in the British Museum.

More recently, we know that it was Halley's Comet in 1066 that tolled the knell for King Harold at the Battle of Hastings. Its depiction on the Bayeux

Halley's Comet appeared in 1066, before the Battle of Hastings, and is shown on the Bayeux Tapestry.

The Italian artist Giotto di Bondone depicted a comet in the sky in the 'Adoration of the Magi', which he painted after the appearance of Halley's Comet in 1301.

Returns of Halley's Comet

240 BC First recorded sighting.

164 Seen by the Babylonians.

87 Seen by the Babylonians and Chinese.

12 Watched by Chinese for two months. 'Hung like a sword over Rome before the death of Agrippa', according to the historian Dion Cassius.

AD 66 'A comet of the kind called Xiphias, because their tails appear to represent the blade of a sword', was seen above Jerusalem before its fall, according to the Jewish historian Flavius Josephus.

141 Described by the Chinese as bluish-white in colour.

218 Described by the Roman historian Dion Cassius as 'a very fearful star'. Heralded the murder of Emperor Macrinus of Rome by his own troops.

295 Seen in China, but not spectacular.

374 Comet passed 13.5 million kilometres from Earth.

451 Appeared before the defeat of Attila the Hun at the Battle of Chalons.

530 Noted in China and Europe, but not spectacular.

607 Comet passed 13.5 million kilometres from Earth.

684 First known Japanese records of the Comet. Seen in Europe and depicted 800 years later in the *Nuremberg Chronicles*.

760 Seen in China, at the same time as another comet.

837 Closest-ever approach to the Earth (5 million kilometres). Tail stretched halfway across the sky. Appeared as bright as Venus.

912 Seen briefly in China and Japan.

989 Seen in China, Japan and (possibly) Korea.

1066 Appeared before the Battle of Hastings. Shown on Bayeux Tapestry. The Comet passed 16.5 million kilometres from the Earth and was described in Europe as looking like a dragon, with multiple tails. Seen for over two months in China.

1145 Depicted on the Eadwine Psalter, with the remark that such 'hairy stars' appeared rarely, 'and then as a portent'.

1222 Described by Japanese astronomers as being 'as large as the half Moon . . . Its colour was white but its rays were red.'

1301 Seen by Giotto Di Bondone and included in his painting 'The Adoration of the Magi'. Chinese astronomers compared its brilliance with that of the first-magnitude star Procyon.

1378 Passed within 10 degrees of the north celestial pole, closer than at any time uring the past 2000 years. This is the last appearance of the Comet for which Oriental records are better than Western ones.

1456 Observed in Italy by Paolo Toscanelli, who said its head was 'as large as the eye of an ox', with a tail 'fan-shaped like that of a peacock'. Arabs said the tail resembled a Turkish scimitar. Turkish forces attacked Belgrade.

1531 Seen by Peter Apian, who noted that its tail always pointed away from the Sun. This sighting was included in Halley's table.

1607 Seen by Johannes Kepler. This sighting was included in Halley's table.

1682 Seen by Edmond Halley at Islington.

1759 Return predicted by Halley. First seen by Johann Palitzsch on December 25, 1758.

1835 First seen at the Vatican Observatory in August. Studied by John Herschel at the Cape of Good Hope.

1910 Photographed for the first time. Earth passed through Comet's tail on May 20.

1986 Encountered by space probe.

Tapestry, being pointed out to King Harold by his worried aides, is perhaps *the* most famous image of a comet in history. An earlier appearance of Halley's Comet, in AD 684, is illustrated in the *Nuremberg Chronicles*, but this was not published until 1493, and the simple woodcut is an artist's impression made 800 years after the fact. According to the accompanying text, the Comet's appearance was followed by three months of rain and storms, culminating in plague.

Eadwine, a Canterbury monk, sketched the Comet at its 1145 apparition on the bottom of an illuminated manuscript of *Psalms* he was transcribing and hence immortalized himself in a way he did not expect. Among those who marvelled at the Comet on its appearance in 1301 was the Italian artist Giotto di Bondone who depicted it as the Star of Bethlehem, darting golden fires in the background of his fresco of the 'Adoration of the Magi'. Now, the European space probe to Halley's Comet bears Giotto's name.

This Babylonian clay tablet contains observations of Halley's Comet at its appearance in 167 BC.

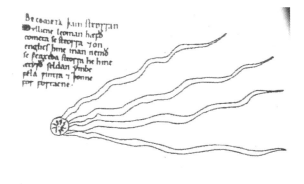

Halley's Comet in 1145, drawn in the margin of the Eadwine Psalter, an illuminated manuscript of the Psalms.

The orbit of Halley's Comet

Halley's Comet takes an average of 76 years to orbit the Sun, covering 11 500 million kilometres in the process, but in practice the periods can be a few years longer or shorter than this. Measured from one perihelion to the next, its quickest return was after 74.4 years, from November 1835 to April 1910, while the longest interval was 79.25 years, from June 451 to September 530. At its most distant (aphelion), the Comet retreats to 5300 million kilometres from the Sun, beyond the orbit of Neptune. At its closest (perihelion) it passes 88 million kilometres from the Sun, between the orbits of Venus and Mercury. In accordance with the laws of planetary motion, when farthest from the Sun the Comet is moving slowest, at 3280 kilometres per hour, while at its closest it is moving most quickly, at 196 000 kilometres per hour. On each orbit the exact figures differ slightly. One fear can immediately be dispelled: there is no chance of a collision of the Comet with the Earth, or any other planet for that matter. The closest that the Earth comes to the Comet's orbit is 10 million kilometres.

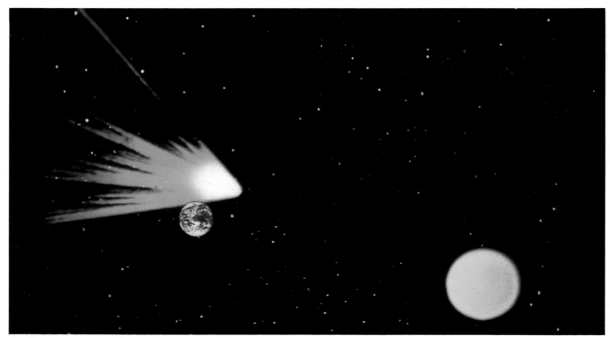

Halley's Comet made its closest-ever approach to the Earth in AD 837, when it passed a mere 5 million kilometres from us.

The orbit of Halley's Comet around the Sun is a long, thin ellipse that is tilted at 18 degrees with respect to the Earth's orbit.

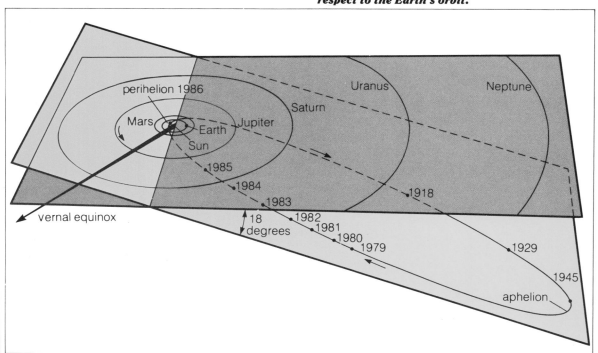

The comet cloud

Beyond the planets, in the darkness far from the Sun, orbits a swarm of 100 000 000 000 comets. No one has actually seen this swarm, for at that distance comets are too faint to detect with any telescope. Rather, the existence of the swarm has been deduced from the fact that virgin comets, making their first approach to the Sun, arrive on highly elongated orbits that must originate partway to the nearest star.

According to the most widely accepted theory, comets are debris remaining from the formation of the Solar System 4600 million years ago. An alternative view is that the Sun sweeps up new batches of comets every so often as it penetrates clouds of gas in the Galaxy. Either way, the comets tag along unseen in cold storage at the perimeter of the Solar System until diverted onto new routes by the gravity of passing stars. Some comets are elbowed out of the Solar System entirely, but others head Sunwards.

Astronomers estimate that the cometary cold store, known as the Oort Cloud after the Dutch astronomer Jan Oort, lies about 50 000 times farther

Ice evaporates from the nucleus of a comet as it approaches the Sun, forming the halo of gas called the coma.

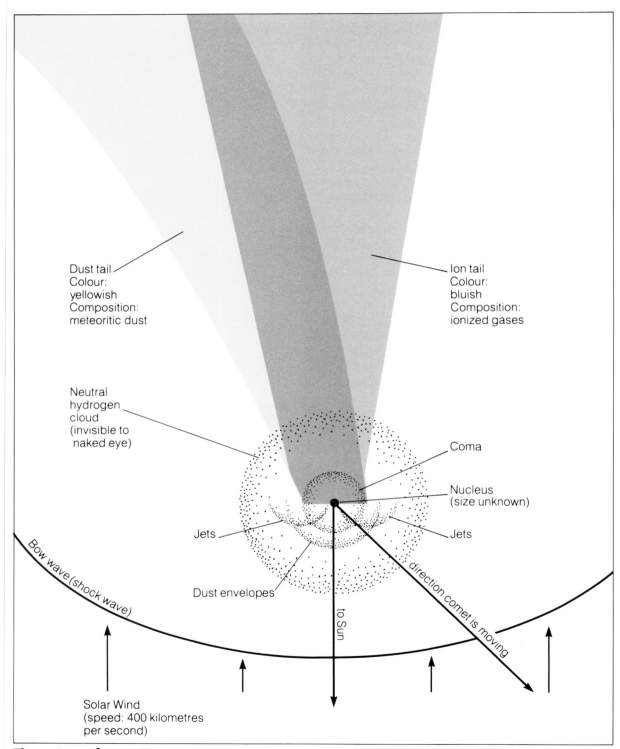

The anatomy of a comet.

from the Sun than does the Earth, a distance that light takes nearly a year to cross. Even that far away the Oort Cloud is still held by the Sun's gravity, albeit loosely. Every few hundred thousand years a star passes close enough to disturb the comets in the cloud onto new orbits. To account for the regular supply of comets heading towards the Sun, astronomers estimate that there must be 100 000 million of them in the cloud. Yet the mass of all these comets in the Oort Cloud is, in total, only a few per cent of the mass of the Earth. Although comets advertise themselves impressively, in truth they are disappointingly insubstantial.

When far from the Sun, a comet resembles a dirty snowball a few kilometres wide. Only when it reaches somewhere around the orbit of Mars do the Sun's light and heat awake the snowball from its hibernation so that it displays itself for the admiration of astronomers.

A comet's regalia has four components: a nucleus; a coma, which together with the nucleus makes up the head; a tail (though not in all cases); and, around the whole comet, an immense bubble of hydrogen, millions of kilometres across, invisible to the eye but detectable by satellites in space.

The fount of all cometary activity is the nucleus, the dirty snowball, composed mostly of frozen water caked with dark dust. Heat from the Sun works its gentle erosion on the nucleus, evaporating the ice to form an enveloping halo of gas and dust, the coma. An ordinary comet's coma is about 10 000 kilometres wide, big enough to engulf the Earth, but in the greatest comets, Halley's among them, the coma spans more than 100 000 kilometres, half the distance from the Earth to the Moon. For all its size, the coma is still transparent – stars can be seen shining through it, as they can through the tail. The gases of a comet are far thinner than your breath.

Comet tails

In the popular mind a comet would not *be* a comet without its tail, but in truth the majority of comets are as tailless as a Manx cat. Amputee comets are almost invariably faint, though, so just about every comet that you are likely to see with the naked eye will have a tail, or possibly two: one made of gas and one of dust. You can tell the two apart because the tail of gas is straighter, whereas the tail of dust is more curved and smudgy. Note that the direction in which a comet spreads its tail is no clue to the course the comet is steering, for a tail is not the wake of a comet. Rather, the tail blows away from the Sun, so a departing comet actually chases its own tail.

Gas tails of comets are evidence that there is a 'wind' of sorts in space, known as the Solar Wind. It is a stream of atomic particles from the Sun which blows gas away from the comet's head like a flag in a breeze. Bombarded by the Sun's ultraviolet light, the gas in the tail fluoresces like an advertising sign, as does the gas in the coma, producing an ethereal self-luminance. Gas tails can extend 100 million kilometres or more, as did that of Halley's Comet in 1910.

Dust tails are shorter, and shine simply by reflecting sunlight. Impacts by photons of sunlight push dust specks away from the comet's head, an effect known as radiation pressure. As the dust recedes from the head it spreads out into a fan shape.

Clearly, a comet pays for its moment of glory when near the Sun by irretrievably shedding mass – a hundred million tons each trip, in the case of a bright comet like Halley's. Dr David Hughes of the University of Sheffield has calculated that the nucleus of Halley's Comet loses a surface layer 2 metres deep at each return. According to Hughes, the nucleus is now 9 kilometres in diameter, about half the size it was when it first approached the Sun 170 000 years ago. Since then, it has made about 2300 returns. Yet the nucleus is still big enough to survive for a similar length of time in the future before it finally fades out. Halley's Comet is in comfortable middle age.

Cometary fireworks

Even if you never view Halley's Comet itself, you can see bits of it twice every year, in May and October. At those times, the Earth shoulders its way past the trail of dust laid by the Comet on its travels around the Solar System, and on each occasion astronomers go out to watch a shower of shooting stars.

A shooting star – really no star at all – is a brief flash of light caused by a speck of cometary dust burning up high in the atmosphere. Astronomers term them meteors. Watch for an hour on any clear night and you will see a few stray meteors, long since separated from their parent comets, plunging to their fiery deaths 100 kilometres above your head. But several times a year the Earth passes close to a comet's orbit, and encounters a dust storm which produces a celestial firework display known as a meteor shower.

Dust spread along the orbit of Halley's Comet causes two meteor showers each year: one in the third week of October, when we pass 23 million kilometres from the inbound leg of the comet's orbit, and the other in the first week of May, when we pass 10 million kilometres from its outbound track. To an astronomer, a 'shower' means perhaps one bright meteor every

Comets shed dust along their orbits. When the Earth meets this dust a meteor shower results. The dust from Halley's Comet causes two meteor showers each year, in May and October, even though the Comet itself may be nowhere in sight.

All the members of a meteor shower appear to diverge from one part of the sky, called the radiant.

five minutes, so do not expect to see the heavens falling. We only dip into the outskirts of the Halley dust storm, so its fireworks are not as spectacular as the displays caused by other comets during the year.

According to ancient superstition, the appearance of a comet heralded the occurrence of diseases that included the plague, as shown in these old drawings. Now, two scientists say that comets may actually bring disease to Earth.

Do comets carry disease?

Professors Fred Hoyle and Chandra Wickramasinghe are present-day astronomers whose ideas about comets have a distinctly mediaeval ring. For, they say, comets are unhealthy bodies that intermittently spray the Earth with germs. Superstitious folk who regarded comets as harbingers of plague and other ailments were not so far from wrong, if Hoyle and Wickramasinghe are to be believed.

These two British scientists think that micro-organisms, including bacteria and viruses that cause disease, can arise within comets. These micro-organisms are sloughed off along with the dust from comets, to be swept up by the Earth from time to time on its journey around the Sun. Among the diseases that Hoyle and Wickramasinghe attribute to comets are plague, smallpox, measles and Legionnaire's disease.

Halley's Comet, they think, carries influenza, for major outbreaks of varieties such as the so-called Asian 'flu seem to recur at roughly 76-year intervals that match the orbital period of the Comet. However, since the Earth encounters dust from Halley's Comet in May and October each year, it is difficult to understand why we are not stricken by Asian 'flu perenially. Few scientists take Hoyle and Wickramasinghe's ideas seriously. If you sneeze when you see a comet, the reason is likely to lie closer to home.

Great comets

Japanese comet hunters have led the world in recent years, among them Kaoru Ikeya and Tsutomu Seki who jointly discovered the celebrated Comet Ikeya–Seki, one of the Sungrazer family. In October 1965, Comet Ikeya–Seki passed less than 500 000 kilometres from the Sun, and its tail was a glorious sight in the morning sky, stretching for over 100 million kilometres. During its closest approach to the Sun, its nucleus seemed to split up, although the comet survived.

Two recent comets demonstrated the unpredictability of cometary behaviour. Both were discovered by chance by professional astronomers on photographic plates taken for other purposes. The first was found by Lubos Kohoutek at Hamburg Observatory in 1973. Calculations of its orbit showed that it would pass 20 million kilometres from the Sun, half the distance of the closest planet, Mercury, and if it brightened as expected it could become the most brilliant comet ever seen. Expectations ran high among astronomers and public alike. Elaborate scientific investigations were planned, astronauts aboard the Skylab space station were alerted to look out for it, and doomsayers proclaimed the End of the World.

In the end, Comet Kohoutek did not brighten by anywhere near as much as had been hoped. It nevertheless proved to be a more than respectable comet, visible to the naked eye in dark skies, and it taught astronomers a lot. But since it did not set the skies ablaze it was widely regarded by the public as a flop.

Not so Comet West of 1976. Discovered by Richard West of the European Southern Observatory in Chile, it proved to be the most spectacular comet of the century. As the comet approached its closest to the Sun, 30 million kilometres, its head became bright enough to see with the naked eye before the Sun had set. After rounding the Sun, Comet West displayed a huge fan-shaped tail with many individual streamers. At the same time, its nucleus broke up into four parts, one of which rapidly expired. Of the others, two departed on orbits that will take them out of the Solar System, while the main fragment of the nucleus will not return to the Sun for half a million years. Alas, Comet West will never again be as spectacular as in 1976.

Naming a comet

Comet hunters are lured by the prospect of seeing some small piece of Solar-System real estate before anyone else. If they succeed, they have it named after them. Sightings must be reported for ratification to the International Astronomical Union's Central Bureau for Astronomical Telegrams in Cambridge, Massachusetts. Up to three names are allowed for comets that have been independently reported by more than one person, as in the case of Comet IRAS–Araki–Alcock. A comet hunter may therefore find his name orbiting the Sun forever, attached to a comet, in conjunction with someone he has never met nor previously heard of.

In addition to the discoverer's name, a comet is allocated a temporary number. After the comet has passed perihelion it is given a permanent number. For example, at its last return Halley's Comet was initially tagged 1909c because it was the third comet sighted in 1909. Later it was designated 1910 II, because it was the second comet to pass perihelion in 1910. Sometimes, a comet is named not after its discoverer but after the person who calculates its orbit, as with Comet Halley and Comet Encke.

Comet Ikeya–Seki in November 1965 after it had 'grazed' the surface of the Sun.

Charles Messier, the great French comet discoverer, made a list of fuzzy-looking objects that could be mistaken for comets. Number 1 on his list is the Crab Nebula.

Caroline Herschel discovered eight comets, in addition to helping her brother William to observe the heavens with his large telescopes.

Comet hunters

Every clear night for the past 30 years, an English schoolmaster has scanned the sky with binoculars from the back garden of his home in Peterborough. In this way George Alcock has discovered a total of five comets, all of which now bear his name – immortality written in the sky. His most spectacular discovery was his fifth, in 1983. This comet was also reported independently by a Japanese amateur and, from space, by the Infra-Red Astronomy Satellite, IRAS, so it was given the cumbersome title of Comet IRAS–Araki–Alcock.

On May 11, 1983, this comet passed 4.5 million kilometres from the Earth, closer than any other except Lexell's Comet, which missed the Earth by a mere 1.2 million kilometres (only three times the distance of the Moon) in 1770. On the night of its closest approach, Comet IRAS–Araki–Alcock was visible like a smudgy thumbprint several times larger than the Moon, though without a tail. Radio astronomers in the USA succeeded in bouncing radio waves off its nucleus, which they concluded was rough, lemon-shaped and 6 to 8 kilometres across.

Dedicated comet hunters such as George Alcock continue a tradition that goes back two centuries. First of the great comet hunters was the Frenchman Charles Messier. Inspired by the return of Halley's Comet in 1758, Messier went on to bag a total of 21 comets of his own, earning the nickname from Louis XV of the 'comet ferret'. But even he was outdone by another Frenchman, Jean Louis Pons, who has three dozen comets to his name, more than anyone else. One of them he discovered twice, in 1805 and 1818. Shortly thereafter the German astronomer Johann Encke calculated that these sightings were of the same comet orbiting the Sun every 3.3 years, the shortest period known, and the object has now been renamed Comet Encke. Unfortunately, Encke's Comet has been Sunblasted so much that it is faint and uninspiring.

Last century, substantial prizes were offered to the discoverers of comets, and one particularly successful American comet hunter, Edward E. Barnard, paid off the mortgage on his house with his winnings. Since then, photography with large telescopes has provided a powerful weapon for comet spotting, but there is still room for the amateur with simple equipment.

George Alcock settles down for a night of comet hunting in his back garden.

Comet IRAS–Araki–Alcock, observed by the IRAS satellite and colour-coded to show regions of different temperature.

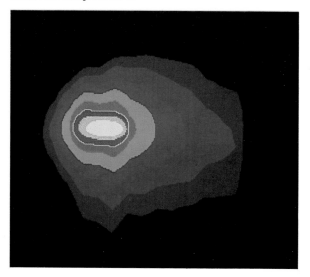

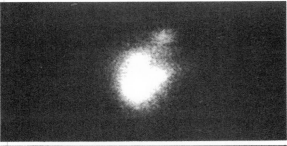
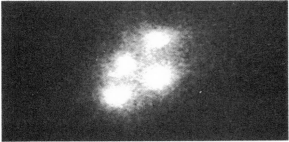
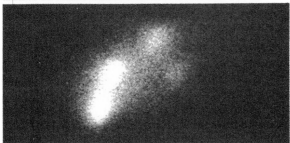
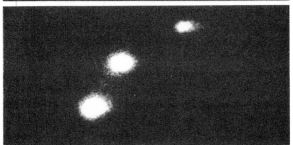

The nucleus of Comet West broke up into four pieces in March 1976.

The return of 1910

Before Comet Halley's appearance in 1910, the German Astronomical Association, the *Astronomische Gesellschaft*, offered a prize of 1000 Marks (then worth £50) for the best prediction of the Comet's forthcoming passage past the Sun. To ensure impartiality in judging, entries were to be sent in anonymously, with the name of the author sealed in an accompanying envelope. Each entry was to be identified only by a motto at its head.

At the Royal Observatory in Greenwich, Philip H. Cowell and Andrew C.D. Crommelin took up the German challenge. They predicted a return to perihelion on April 16, 1910. They sent in their entry to the *Astronomische Gesellschaft* under the motto *Isti Mirantur Stellam*, 'They marvel at the star', the legend on the Bayeux Tapestry.

Observatories around the world raced each other to sight the returning Comet. Max Wolf of Heidelberg spotted it first, on a photograph he took early on the morning of September 12, 1909. Now the predictions could be judged against the actual path of the Comet. When the sealed envelopes were opened, the winners were Cowell and Crommelin.

Not content, the two Greenwich astronomers continued to refine their calculations, taking the gravitational effects of all the planets into account, and published their results in an 84-page book. Their revised date of perihelion was for early on April 17. But, despite their best efforts, the Comet actually got to perihelion three days later than they predicted. Cowell and Crommelin concluded: 'There are forces of an unrecognized kind influencing the comet's motion'.

What those mysterious forces were was not known for certain until 1950, when they were explained by Fred Whipple, progenitor of the dirty snowball theory of comets. Puffs of gas evaporating from the nucleus produce a rocket effect, like the manoeuvring jets of a spacecraft, which pushes the Comet slightly off course. Astronomers term these effects non-gravitational forces, and they now take them into account when calculating the path of a comet.

Awaiting the Comet

Public anticipation of the Comet's reappearance was immense. Tunesmiths composed songs to serenade the heavenly visitor, and poets burst into verse. Products such as Bird's Custard and Pears' Soap featured the Comet in their advertising: 'Pears' Soap is visible day and night all over the world', was one slogan. Even before the Comet was visible to the naked eye, people wrote to the Royal Observatory to report their sightings, which turned out to be misidentifications of the bright planets Venus and Jupiter and in one case the Andromeda Galaxy.

For a while, though, it looked as if Halley's Comet would be upstaged by a pretender: a *second* comet, which appeared unexpectedly in the dawn sky in January 1910, and was first seen by miners in Johannesburg leaving night shift. For a few days it became visible in daylight and it is known as the Daylight Comet of 1910. But it had faded from view by the time Halley's Comet bowed into the morning sky in mid-April.

To an observer in Accra, West Africa, Halley's Comet appeared 'like a flaming sword with jewelled hilt'. Over the next few weeks its tail stretched upwards like a searchlight beam illuminating the roof of the heavens. In early May, Halley's Comet lay near the brilliant planet Venus, and superstitious Englishmen marked that this coincided with the death of King Edward VII.

When the public heard that the Earth was due to pass through the Comet's tail, there was widespread disquiet. A family contemplating a sea journey inquired of the Royal Observatory whether the Comet would cause storms and gales 'or other violent atmospheric disturbances'. They were assured not.

A correspondent with a taste for apocalyptic prose confided to the Observatory his suspicion that the Comet's tail, in contact with the atmosphere, would 'cause the Pacific to change basins with the Atlantic, and the primeval forests of North and South America to be swept by the briny avalanche over the sandy plains of the great Sahara, tumbling over and over with houses, ships, sharks, whales and all sorts of living things in one heterogeneous mass of chaotic confusion'. The Observatory tersely marked the letter 'No reply'.

Other reactions ranged from the light-hearted to the grotesque. From Paris it was reported that restaurateurs were preparing comet suppers for the

Science Jottings—By "Dr." W. Heath Robinson (D— L—).

Cartoonist Heath Robinson's view of the search for Halley's Comet at Greenwich Observatory in 1910.

great occasion, and comet postcards and souvenirs were selling well. In the USA, churches were packed with people who feared that the encounter with the Comet signalled the end of the world. A shepherd in Washington State was reported to have gone insane with worry about the Comet, while in California a prospector nailed his feet and one hand to a cross and, despite his agony, pleaded with rescuers to let him remain there.

Cyanogen gas, a poison, had been discovered spectroscopically in the tail of Comet Morehouse in 1908, so people feared, understandably, that they might be poisoned by gases from the tail of Halley's Comet. From Chicago it was reported that women were stopping up doors and windows to keep out the toxic vapour. In Haiti a voodoo doctor sold comet pills to ward off the evil influence of the Comet, as did two swindlers in Texas who also did a good trade in leather gas masks. Purchasers were told that the pills (actually made of a harmless combination of sugar and quinine) would help them withstand the gases of the Comet's tail. Police arrested the men but were forced to let them go again when their gullible victims campaigned for their release.

Astronomers tried to reassure the public that there was no danger. For one thing, the head of the Comet would come no closer to us than 24 million kilometres. Some scientists thought that a meteor shower or an aurora was possible, but most held to the opinion that there would be no observable effects at all. In fact, the Earth had passed through a comet's tail at least once before, that of Comet Tebbutt in June 1861, and nobody noticed a thing.

Through the Comet's tail

About 3 a.m. GMT on May 19, 1910, Halley's Comet passed directly between the Sun and the Earth. This event was invisible from Greenwich, the Sun being below the horizon, but observers on the other side of the world, in Hawaii, trained their telescopes on the Sun for signs of the Comet's head silhouetted against the Sun's disk. They saw nothing. Had there been a solid nucleus as little as 100 kilometres across, the astronomers would have seen it as a dark dot crossing the Sun.

Those who believed that the Earth's passage through the Comet's tail would mark the end of the world must have feared the worst when violent thunderstorms broke out over England that night. From Leigh-on-Sea in Essex, an imaginative observer described the lightning as being 'almost the colour of blood'. At the Paris Observatory, Camille Flammarion reported that four observers 'had certain olfactory experiences, which are described variously as a smell of burning vegetables, of a marsh, or of acetylene . . .'. Imagination must have got the better of them, for the Earth's atmosphere would have prevented the thin gases of the Comet from penetrating any closer than about 100 kilometres from the ground.

From Greenwich on the night of the Earth's passage through the tail, Crommelin noticed strange bands of light in the sky. At first he put them down to clouds but later he wondered whether they were any-

Halley's Comet appeared close to Venus in the morning sky in May 1910, when the Comet's tail streamed out for 100 million kilometres.

thing to do with the Comet. The Engineer-in-Chief of the General Post Office wrote to the Astronomer Royal to inform him that no electrical effects were noted on telephone trunk lines during the Earth's passage through the tail. With hindsight, it now seems that the Earth did not pass through the centre of the tail, but only through its outskirts.

Perhaps the strangest letter about the encounter to be received at Greenwich came from Sze zuk Chang Chin-liang, who wrote from the Imperial Polytechnic College, Shanghai. He thoughtfully enclosed a photograph of himself to accompany his revolutionary theory: 'It is obvious the comet has no tail at all and the so-called tail must be the Sun rays which, while passing through the body of the comet, look like a tail'. He then confessed his fear: 'If the body of the comet is transparent and like the Earth has its two poles fairly flat and thus form a convex lens then everything on the Earth will be burnt provided the sunlight passes through the body of the comet and the focus falls on the surface of the Earth'. Since this letter is dated 1912, two years after the event, it seems rather late to have worried.

After its passage across the face of the Sun on May 19, Halley's Comet reappeared in the evening sky, still impressive but fading as it receded. By July it was lost to the naked eye. Telescopes followed the Comet for a year until it was more distant than Jupiter. Then it was gone from sight. Halley's Comet was not seen again for another 71 years.

Giotto, the European space probe to Halley's Comet, launched by an Ariane rocket on July 2, 1985.

Space probes to Comet Halley

Halley's Comet was the obvious candidate for the first space mission to a comet: it is big and bright, and its return could be predicted years in advance. Lack of money prevented the American space agency, NASA, from launching a probe of its own. That left the field open for the rest of the world. Five spacecraft are being sent – two from the Soviet Union, two from Japan, and one from Europe. They will all meet Halley's Comet in March 1986, when it will be at its brightest after perihelion.

NASA will, however, launch a Space Shuttle mission, called Astro-1, to observe the Comet from orbit. Also, a NASA spacecraft for studying the Solar Wind has been diverted to fly through the tail of a much fainter comet, Giacobini–Zinner, in September 1985. That spacecraft, now named ICE (International Cometary Explorer), will pass between the Sun and Halley's Comet in October 1985 and March 1986, measuring the Solar Wind blowing towards the Comet. ICE will not fly close enough to study Halley's Comet itself.

Japan

MS-T5, also known as *Sakigake* (meaning Pioneer), is a pathfinder probe launched by Japan on January 7, 1985. It will pass 1 million kilometres from Halley's Comet in March 1986. The main Japanese probe, Planet-A, was due to be launched on August 14, 1985, to take photographs in ultraviolet light of the enormous cloud of hydrogen gas, several million kilometres across, surrounding the Comet. At its closest to the Comet, on March 8, 1986, Planet-A will be inside the hydrogen cloud.

Soviet Union

First to depart for Halley's Comet, in December 1984, were the Soviet probes Vega 1 and 2. Their name is short for Venus–Halley (in Russian). They dropped off two landers at Venus in June 1985 before proceeding to the Comet. Vega 1 will pass 10 000 kilometres from the Sunward side of Halley's nucleus on March 6, 1986. Its results will be used to guide Vega 2, due for encounter three days later. Both Vega probes carry cameras to photograph the Comet's head and tail, and instruments to analyse the composition of the gas and dust around the Comet.

Europe

Giotto is the European Space Agency's probe to Halley's Comet, named after the Italian artist who depicted the Comet in his painting of the 'Adoration of the Magi'. It is the most ambitious of all the probes, for it is targetted to fly a mere 500 kilometres from the Comet's nucleus, photographing it in colour. Other instruments will analyse the dust and gas of the Comet's head. If theories about comets are right, we shall be sampling the primordial material from which the Solar System formed, 4600 million years ago. Giotto was launched on July 2, 1985. After a trip of 685 million kilometres, it will meet the Comet on the night of March 13, 1986. Giotto and the Comet will hurtle towards each other at the colossal speed of 240 000 kilometres per hour, 50 times faster than a bullet from a gun. Despite a protective two-layer shield, Giotto will probably be destroyed by impacts with dust particles as it plunges towards the nucleus. By then it will have transmitted its findings to Earth.

After an eight-month journey from Earth, Giotto will pass through the head of Halley's Comet on the night of March 13, 1986.

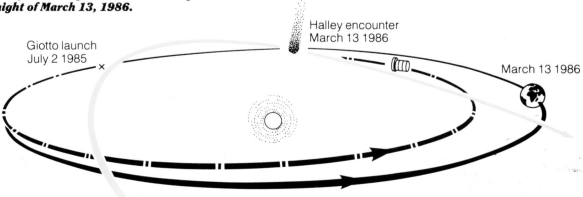

Giotto launch
July 2 1985

Halley encounter
March 13 1986

March 13 1986

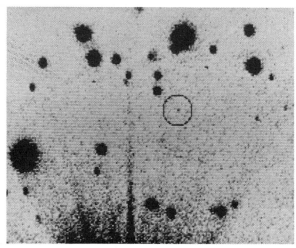

The Japanese probe Planet-A.

Recovery photograph of Halley's Comet, taken at Palomar Observatory on October 16, 1982. The Comet is the tiny dot within the circle.

(left) Two Soviet probes called Vega are leading the way to Halley's Comet. Their photographs will help controllers to steer the European probe Giotto as close as possible to the Comet's nucleus.

Looking for Halley's Comet

On October 16, 1982, Halley's Comet came within reach of telescopes on Earth for the first time since 1911. Two American astronomers, David Jewitt and G. Edward Danielson, sighted it with a specially sensitive detector attached to the 5-metre reflector on Mount Palomar. At that time, Halley's Comet lay beyond the orbit of Saturn and was as faint as a candle 43 000 kilometres away. To the relief of those planning space probes, the Comet was following closely the path predicted by Donald Yeomans of the Jet Propulsion Laboratory, California, that would bring it to perihelion on February 9, 1986.

Halley's Comet will be visible using small instruments and to the naked eye for two to three months before and after perihelion. If it behaves as on previous appearances, it should be significantly brighter after perihelion than before. Astronomers throughout the world will be studying the Comet intently as part of a coordinated programme called the International Halley Watch.

Many people were disappointed by the Comet in 1910 because they tried to observe it from amid the smoke and artificial light of towns. Think how much worse things have got in the past three-quarters of a century. The shy glow of comets is easily swamped by other sources of light, including moonlight, so to get the best views you must go where the air is clear and dark, well away from towns. Remember also that the Comet will be much more difficult to spot when it is close to the horizon, because it is then dimmed by the Earth's atmosphere.

Your location on Earth will affect what you see of the Comet. The further south you are, the better. When at its brightest, in spring 1986, the comet will be too far south to see from northerly latitudes including most of Europe and the northern parts of North America.

Nevertheless there will be an opportunity for everyone, aided by a pair of binoculars, to view this once-in-a-lifetime celestial visitor. You will be watching an object that was seen by Julius Caesar in 87 BC, King Harold in 1066, Genghis Khan in 1222, Giotto in 1301, William Shakespeare in 1607 and, of course, Edmond Halley in 1682.

Where to see Halley's Comet

As Comet Halley approaches the Sun in the autumn of 1985 it will lie in the northern half of the sky. By the middle of November 1985 it will be moving through the stars of Taurus and will have brightened sufficiently to be visible with small telescopes and binoculars, though not yet to the naked eye. Taurus is high in the south-east by 10 p.m. On the night of November 16, the Comet will lie 2 degrees (four Moon diameters) below the Pleiades star cluster in Taurus. Through binoculars, the Comet should be visible in the same field of view as the Pleiades. It will appear like an elongated smudge. This is the time to start looking for the Comet, as there will be no moonlight to interfere.

By mid-December the Comet will have brightened to the verge of naked-eye visibility, for those in the darkest skies. It will certainly be easy to see through binoculars, most probably showing a tail. Look for the Comet from December 6 to 15 when it will lie below the Square of Pegasus, due south at 6 p.m. for northern-hemisphere observers.

During December the Comet will move south- wards, taking it lower in the sky, the tail developing as it approaches the Sun. On Christmas Eve the Comet will cross the celestial equator, setting by 10 p.m. The last few days of 1985 and the first week of 1986 are probably the best times for observers above latitude 50 degrees north to sight Halley's Comet. It will lie in the south-west for a few hours after sunset, in the constellation Aquarius. As the Comet will not yet be prominent to the naked eye, the best views will still be gained through binoculars.

January 1986 should see the Comet moving further south and closer to the Sun, vanishing in the evening twilight after the middle of the month. Following perihelion, which will occur on February 9, the Comet will emerge into the dawn sky, though not until the last week of February. Now will come the best part of the Halley show, though you will need to be situated south of latitude 40 degrees north to appreciate it.

Halley's Comet should now be a naked-eye object, at least for those observing away from towns. During March the Comet will approach Earth in the morning sky, its tail growing so that it will appear longer than your hand held at arm's length. On March 13, when

The path of Comet Halley 1984-1986

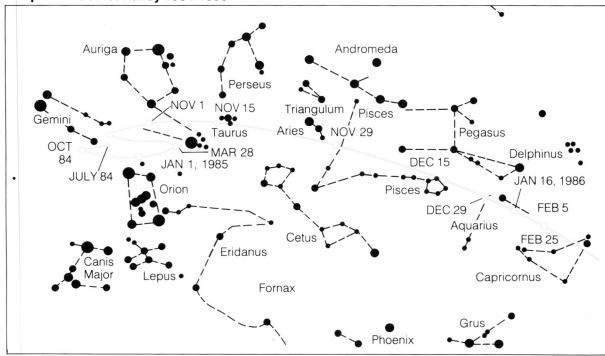

Giotto reaches it, the Comet will be on the border between Capricornus and Sagittarius, a beautiful ghostly object in the east before sunrise.

Towards the end of March the Comet will back rapidly away from the Sun, rising earlier all the time, but also dropping further south. Unfortunately for observers at 50 degrees north it will sink below the southern horizon by the beginning of April, not to reappear for another two weeks, and during this time it will also be too low to be seen from 40 degrees north. But in the southern hemisphere Halley's Comet will be riding high.

Moonlight will interfere with the Comet's visibility in early April, so the second week of April is the time to travel south to admire the Comet's splendour. Circumstances will be ideal on April 11, when the Comet is closest to Earth and there is no moonlight. By then, Sun and Comet will be almost opposite each other in the sky so that the Comet is again an evening object rising at about sunset. From latitude 30 degrees south the Comet will be high in the sky at 10 p.m., among the stars of Lupus.

If it has lived up to expectations, Halley's Comet should appear like a moderately bright star out of focus, with a head perhaps half the size of the Moon and a prominent tail, reminiscent of the old Chinese descriptions. Compare it over the next few nights with Omega Centauri, a rounded star cluster of similar apparent size and brightness to the Comet's head, which lies a few degrees away. Look for changes in the tail from night to night. On April 24 a total eclipse of the Moon will darken the sky in western North America, Australasia, Japan and the Pacific, affording an additional chance to see the Comet.

Once it has passed Earth, the Halley show will rapidly wane. The Comet's brightness will drop as it recedes from the Sun and Earth; but the Comet will climb northwards again to bring it above the horizon for observers at latitude 50 degrees north in the third week of April. It will make its reappearance in the evening sky, slightly brighter than before perihelion, though difficult to spot because it is low down.

As the Comet moves northwards it will fade, and by the end of May it will no longer be visible without binoculars. Large telescopes will continue to follow Halley's Comet for as long as possible as it speeds away into the darkness, not to return to perihelion again until July 2061.

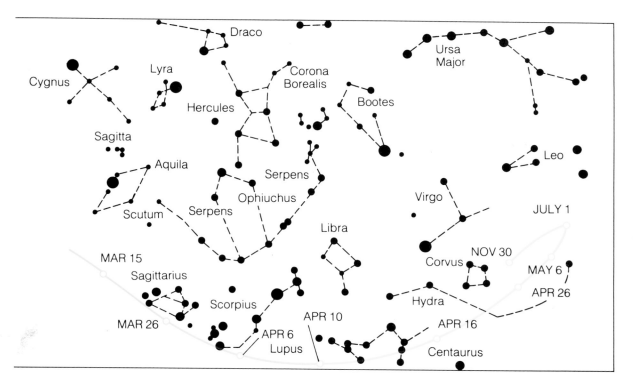

43

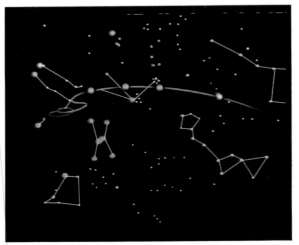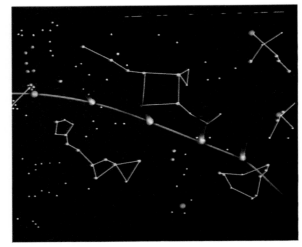

These four charts show the path of Halley's Comet from 1982 until 1986. They were drawn on a Quantel 'Paintbox' computer graphics system and for ease of understanding should be compared to the main star chart (pages 42 and 43).

Looking from left to right, the first picture shows the path of the Comet since it was first detected in 1982, through to December of 1985 by which time the Comet should have reached naked-eye visibility.

Next, we follow the Comet through perihelion, its closest approach to the

How to observe Halley's Comet

From what has been said above, it is clear that a pair of binoculars is essential to view Halley's Comet. Binoculars are better than telescopes because they are cheaper, portable and have a wider field of view, which is important when dealing with an extended object like a comet. Even when the Comet is visible to the naked eye, binoculars will be useful for picking out faint detail in the tail and the head.

Binoculars carry markings such as 8 × 30, 8 × 40, 7 × 50 and 10 × 50. The first figure is the magnification, and the second figure is the width of the front lens in millimetres – the larger this lens the better, for it will show fainter objects. Almost any binoculars are suitable for watching the Comet, but beware of cheap binoculars that offer very high powers, as these are often of poor optical quality. In any case, high magnifications are not necessary. Rather, for spotting hazy objects like comets you need a low magnification to make the comets appear brighter against the sky, and this applies when observing with telescopes as well.

Do not spend a lot of money on a telescope specially to see Halley's Comet, unless you are going to take up astronomy seriously. Small telescopes will show the comet no better than a pair of binoculars, yet they will be much more expensive. A large telescope is a precision optical instrument and costs as much as a hi-fi or a top-quality camera. Incidentally, telescopes used for astronomy show the image upside down, although you can also buy interchangeable eyepieces that turn images the right way up for normal viewing on Earth.

Whether you view with optical equipment or not, you will find the Comet easier to see if you look slightly to one side of it, so that its light falls on the outer part of your retina, which is more sensitive. This technique is called averted vision, and is useful when viewing any elusive objects. Remember also that your eyes need time to grow accustomed to the dark when you first go outside from a bright room. For your eyes to become properly dark adapted takes at least 10 minutes, and the longer you stay out, the more you will be able to see of the Comet's faint details. Be sure to wrap up warmly for your Halley watches.

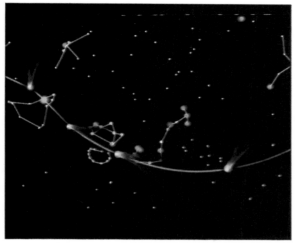 

Sun, and through to March when it becomes increasingly visible with possibly two tails.

The third chart shows the path of the Comet as it plunges Southwards in the sky when it will be at its best for viewing from Earth, having made its closest approach to our world in early April.

The final chart shows how the Comet moves through the end of April and into May. By the end of May the Comet will be exceedingly faint and for most amateur astronomers this will be their last glimpse of Halley's Comet.

How to photograph Halley's Comet

If you can see Halley's Comet with the naked eye, you can photograph it, either in colour or in black and white. You will need a camera with a shutter that can be held open to take a time exposure. Mount the camera firmly on a tripod or wedge it in position with wood or bricks so that it will not move during the exposure. You will need a cable release that can be locked to keep the shutter open. Focus the camera on infinity, and set the lens to its widest aperture.

Which film should you choose? If you process your own photographs, you might want to use fast black-and-white film. For colour, slides are preferable to prints. Use one of the very fast colour films now available, rated 1000 or 1600 ISO (formerly ASA).

Take a series of exposures of different lengths – say, 15, 20, 30 and 40 seconds. Longer exposures will mean that the rotation of the Earth will blur the stars and Comet. On a photograph exposed for several minutes, images of the stars and Comet will be drawn out into long trails by the Earth's rotation which will spoil your picture. One useful tip is to place a piece of card over the camera lens while you open the shutter, in case you accidentally jog the camera. Remove the card for the required exposure time, then replace it in front of the lens before closing the shutter.

You can also try different lenses. A wide-angle lens will show a wider area of sky, and some foreground to make the picture more attractive, but the Comet will appear smaller. A telephoto lens will give a bigger image of the Comet, but it will also magnify the effect of the Earth's rotation, so you will need to keep your exposures shorter – no more than 10 seconds for a 135-millimetre lens. All told, the standard lens may be the best compromise. Try taking a series of test shots and have them processed before the Comet arrives. That way you will learn what works best, and you will get some interesting star shots along the way. Write down the details of each exposure as you take it.

Take a whole series of pictures of the Comet on succeeding nights. Remember, you won't get another chance! One final tip: put one or two ordinary daytime scenes on the start of the roll, so that the processors know where to cut the frames. Alternatively, ask for the film to be returned uncut and unmounted.

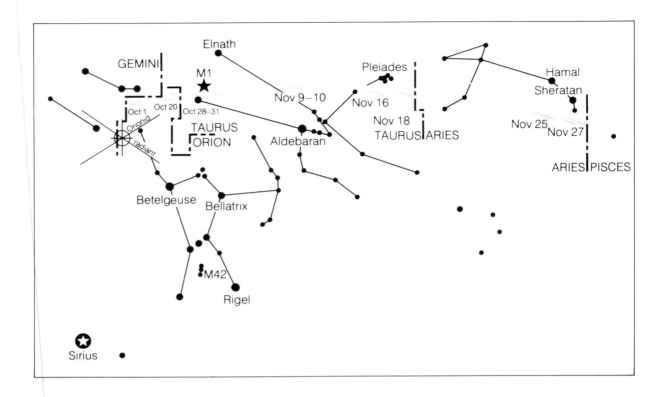

Halley's Comet timetable

1985

October 1   Halley's Comet 305 million kilometres from Earth, 350 million kilometres from the Sun. Comet's speed in orbit 96 000 kilometres per hour.

November 1   Halley's Comet 160 million kilometres from Earth, 287 million kilometres from the Sun. Comet's speed in orbit 106 000 kilometres per hour.

November 27   Halley's Comet makes its closest approach to Earth on its inbound journey, at a distance of 93 million kilometres from Earth (but still 230 million kilometres from the Sun).

December 1   Halley's Comet 94 million kilometres from Earth, 220 million kilometres from the Sun. Comet's speed in orbit 122 000 kilometres per hour. Halley's Comet should become visible to the naked eye this month, in dark skies away from street lights.

1986

January 1   Halley's Comet 173 million kilometres from Earth, 151 million kilometres from the Sun. Comet's speed in orbit 149 000 kilometres per hour. The Comet will become progressively more difficult to see this month, becoming lost in the evening twilight by about the third week.

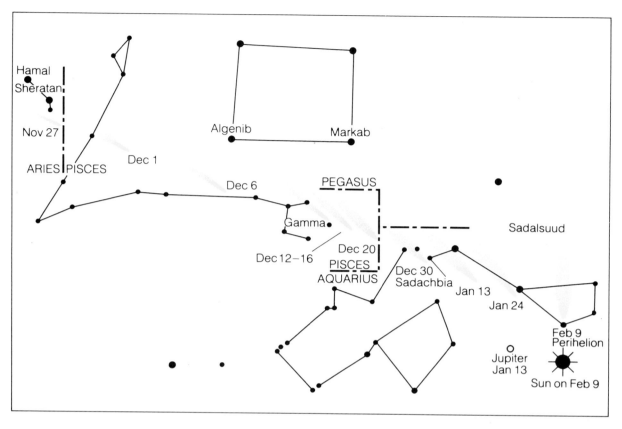

February 1 — Halley's Comet 233 million kilometres from Earth, 93 million kilometres from the Sun. Comet's speed in orbit 191 000 kilometres per hour.

February 9 — Halley's Comet at its closest to the Sun (perihelion), 87.8 million kilometres from the Sun. But it will be invisible, being on the far side of the Sun from Earth (distance from Earth: 232 million kilometres). Comet's speed in orbit 196 000 kilometres per hour.

March 1 — Halley's Comet 190 million kilometres from Earth, 108 million kilometres from the Sun. Comet's speed in orbit 177 000 kilometres per hour. The Comet has now reappeared in the morning sky.

April 1 — Halley's Comet 79 million kilometres from Earth, 176 million kilometres from the Sun. Comet's speed in orbit 137 000 kilometres per hour. The first two weeks of this month are the best time to see the Comet. Best observing sites are in the southern hemisphere, where it will appear overhead.

April 11 — Halley's Comet closest to the Earth on its outbound leg, 63 million kilometres from Earth, but 199 million kilometres from the Sun.

May 1 — Halley's Comet 120 million kilometres from Earth, 244 million kilometres from the Sun. Comet's speed in orbit 116 000 kilometres per hour. The Comet should fade out of sight to the naked eye during this month.

Acknowledgements

We are grateful to the following people and
organisations for permission to use their
illustrations.
Royal Astronomical Society (Comet Burnham p.7,
Comet West p.31)
Lowell Observatory (Comet West p.11)
Royal Greenwich Observatory (Halley's notebook
p.18, Hevelius comet tails p.15, Heath Robinson
cartoon p.35, Halley's Comet.) p.37
British Aerospace (Adoration of the Magi p.21)
British Museum (Babylonian Tablet p.23)
NASA (IRAS-Araki-Alcock comet p.33)
ESA (Giotto probe p.38)

About the authors:

Ian Ridpath is an author and broadcaster on
astronomy and space. His other books include the
Collins Guide to Stars and Planets. He ran the 1985
London Marathon dressed as Halley's Comet.

Terence Murtagh is Director of the Armagh
Planetarium, an author and broadcaster, and
producer of films about space and astronomy.